ISAAC ASIMOV'S
Library of the Universe

Our
Milky Way
and Other
Galaxies

by Isaac Asimov

Gareth Stevens
London • Milwaukee

A note from the editors: In the United States and other places — including this book — a billion is the number represented by 1 followed by nine zeroes — 1,000,000,000. In other countries, including Britain, this number is often called 'a thousand million', and one billion would then be represented by 1 followed by 12 zeroes — 1,000,000,000,000: a million million, which is called a 'trillion' in this book.

The reproduction rights to all photographs and illustrations in this book are controlled by the individuals or institutions credited on page 32 and may not be reproduced without their permission.

A Gareth Stevens Children's Books edition. Edited, designed, and produced by

Gareth Stevens, Inc.
7317 West Green Tree Road   Milwaukee, Wisconsin 53223, USA

ISBN 0-8368-7025-5

First published in the United States and Canada by Gareth Stevens, Inc.
First published in the United Kingdom in 1988 by Gareth Stevens Children's Books

Cover painting © Julian Baum

Designer: Laurie Shock
Picture research: Kathy Keller
Artwork commissioning: Kathy Keller and Laurie Shock
Project editor: Mark Sachner

Technical adviser and consulting editor: Greg Walz-Chojnacki

1 2 3 4 5 6 7 8 9 93 92 91 90 89 88

CONTENTS

Introduction

We live in an enormously large place called the Universe. It's only in the last 50 years or so that we've found out how large it really is.

It's only natural that we would want to understand the place in which we live and in the last 50 years, we have developed instruments to help us learn more about our vast home. We have radio telescopes, satellites, probes, and many other things that have told us far more about the Universe than could possibly have been imagined when I was young.

Nowadays, we have seen planets up close. We have learned about quasars and pulsars, about black holes and supernovas. We have learned amazing facts about how the Universe may have come into being and how it may end. Nothing can be more astonishing and more interesting.

In the sky, we see thousands of stars with our unaided eyes alone. With telescopes we see billions more. Our Sun is just one star in all these billions. We have learned the shape of the vast group of stars of which our star is a member. We have learned of other groups far beyond our own. There are billions of such groups of stars, and these groups are called galaxies. In this book, we will give you a glimpse of these galaxies and tell you something about them.

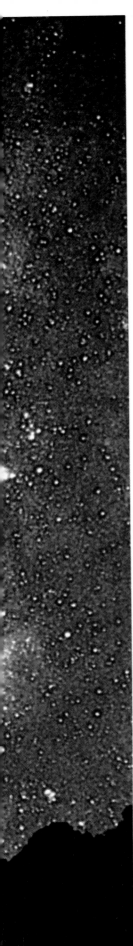

A Milky Glow

If you want an idea of how tiny a part of the Universe is made up of Earth and its human inhabitants, try this: Get out on a dark, clear night, away from city lights, and look up at that faint, foggy band encircling the sky. This band is called the Milky Way. No one knew what it was until the telescope was first used in 1609. Then it was discovered that the Milky Way was a collection of very faint stars — billions of them! Around the year 1800, astronomers decided that the stars exist in a huge collection shaped like a pancake. This collection of stars is called the Galaxy, from a Greek word for 'Milky Way'.

The Milky Way at night. This photograph was taken in the desert in Arizona, USA. It shows what you, too, can see on a clear night if you are out in the country where city lights don't blot out the sky's natural light. Actually, the thick band of light we call the Milky Way consists of billions upon billions of stars that make up one of our Galaxy's spiral arms — the Sagittarius arm.

A portrait of the Milky Way Galaxy: Home, Sweet Home! See
how the four spiral arms swirl out from the centre? Our Sun
and its family of planets — the Solar system — are in the Milky
Way's third arm, which is named for the constellation Orion.

The Milky Way, Our Sun's Home

Astronomers first thought that the Sun must be located near the centre of the Galaxy. Later they found that the centre was 24,000 light-years away from us in the direction of the constellation Sagittarius. By the way, a light-year is the distance travelled by light in one year — nearly 9.5 trillion km, or six trillion miles. At that speed, light travels from the Sun to Earth in about eight minutes. If you could travel at the speed of light, you could go around the world seven and a half times in just one second.

The Galaxy is 100,000 light-years across from end to end. It is made up of a central ball of old reddish stars, and a flat outer disk of gas, dust, and young bluish stars. The Milky Way would appear much brighter near its centre, but dust clouds hide the centre from our view. The Sun is located in the outer disk, in the Orion arm of the Galaxy. The Orion, Centaurus, Sagittarius, and Perseus are the four spiral arms of our Galaxy.

Stars are born out of vast clouds of dust and gas located in the arms of our Galaxy. These clouds are called nebulas, or nebulae. This photo is of a nebula in the Sagittarius arm of our Galaxy as seen from an observatory on Earth.

The centre of the Milky Way. It's not the bright area near the centre of the photo. Rather, it's the darker, curved band just to the <u>right</u> of the brightness. That is where vast clouds of gas and dust have blotted out our Galaxy's brilliant centre.

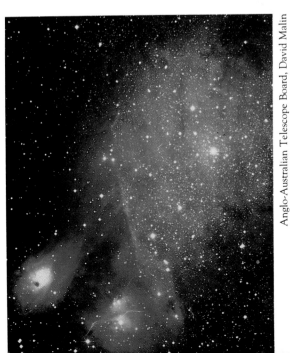

Anglo-Australian Telescope Board, David Malin

© Frank Zullo 1985

The Life and Death of Stars

Stars are what our Galaxy is all about — at least 200 billion stars in all! Stars form out of large clouds of dust and gas. With so many stars in our Galaxy, there are many different kinds of stars at different stages of their lives. Our Sun formed nearly five billion years ago. Other stars are forming today. And since stars don't live forever, many stars are dying at this very moment. The biggest ones eventually explode and add more material to the dust clouds out of which new stars form. Really massive stars live only a few million years before exploding. Our own Sun will live for billions of years before it explodes into a red giant star. It will then come to a quiet end as it shrinks into a tiny white dwarf star.

Here are the stages in the life of a star like our Sun: At left is an accretion disk with a faint glow at its centre. It is forming out of a nebula, which is a huge cloud of gas and dust. To the right, we see the star at its longest stage, as a main sequence star. Our Sun has been at this stage for 4.5 billion years. It will be like this for another 4.5 billion years. Then it will become a red giant for about 500 million years (centre), finally shrinking into a white dwarf (far right). The white dwarf will pack most of the Sun's mass into a body about the size of Earth. It will then spend several billion years cooling off.

At the centre of this photo is a supernova in the mini-galaxy
known as the Large Magellanic Cloud. The largest red giants
explode into supernovas. Our Sun, like most stars, will meet a
less spectacular end as a white dwarf.

How did galaxies form? — will we ever know?

*Astronomers think that when the Universe formed, it was a small
object with all its mass evenly spread out. How did that mass break
up into clumps to form the galaxies? Some astronomers think that
densely-packed matter called black holes formed, and that their
gravitational pull dragged gas and dust about them, and this clumped
together to form stars. That would explain why there are black holes at
the centres of galaxies. Other astronomers have other notions, but no
one really knows.*

Moving into the Big Time — Groups of Stars

The cloud of dust and gas that became our Sun formed only a single star, plus its planets. Such clouds often form more than one star. Double stars — stars that circle each other — are quite common. Some stars consist of two pairs, even three. In fact, stars begin their lives in large groups, and the sky holds many collections of young stars. There are also globular clusters, which are balls of closely packed stars that may number in the hundreds of thousands. And indeed, the whole Galaxy, which is made up of hundreds of billions of stars, probably started as a vast cloud of gas.

© Julian Baum 1988

From the surface of this imaginary planet, you can see a double star system, also called a binary star system. These stars rotate around each other. Here one is much closer than the other. Above them is a globular cluster.

An actual globular cluster. It is 16,000 light-years from Earth —
a close neighbour in globular cluster terms.

© John Foster

A new star cluster would look like
this. There could be hundreds of
stars in this cluster alone.

Our Next-door Neighbours

Does our Galaxy contain all the stars there are? Is it the whole Universe? No, it isn't. In the Southern Hemisphere, one can see two dim clouds in the night sky that look as though they are pieces of the Milky Way which have broken loose. They are called the Large Magellanic Cloud and the Small Magellanic Cloud. They are named after the explorer Ferdinand Magellan, who was the first European to see them. They are also made up of myriads of faint stars. They are 'dwarf galaxies' about 150,000 light-years away. The Large Magellanic Cloud contains about 10 billion stars. The smaller cloud holds only about two billion stars. Compared to them, our Galaxy is a giant!

People south of the Equator can see the Large Magellanic Cloud without instruments. Because it is so close, astronomers have studied this galaxy, which contains about 10 billion stars, to gain insights into our own Galaxy.

NOAO

Also a 'satellite' galaxy of our Milky Way — and also visible to the naked eye south of the Equator — the Small Magellanic Cloud has only about two billion stars. The Milky Way has over 200 billion stars.

NOAO

The rotating galaxies — why won't they behave?

Astronomers have always thought that most of the mass of a galaxy — up to 90 percent — is located in the centre. Stars on the outskirts circle the centre. The farther out they go, the more slowly they move. Astronomers know exactly how the rate of motion should drop off as one moves farther from the centre. When the rotation of galaxies is measured, however, the rate of motion doesn't drop off the way it should. There seems to be more mass in the outer regions than we can make out. And now many scientists think that mass could make up 90 percent of the entire Milky Way — and we're not even sure that mass exists. Perhaps galaxies are surrounded by halos of dim stars or black holes we can't see. Astronomers call this 'the mystery of the missing mass'.

NOAO

The Milky Way's Family

There are still other galaxies. For instance, there is a faint, foggy patch of light in the constellation Andromeda. It had been known for centuries, and for a long time it wasn't considered important. But in the 1920s, astronomers realized it was a distant galaxy, one that was even larger than our own. It is the Andromeda galaxy, and it is 2,300,000 light-years away.

There are about two dozen other galaxies, mostly small ones, that are found at such distances. They make up a cluster of galaxies called the Local Group. All these galaxies are held together by gravity and move about each other.

A spectacular photo of the Andromeda galaxy. Like the Milky Way and over half the other known larger galaxies, the Andromeda is a spiral galaxy. A spiral galaxy is made of a flattened disk at the centre with spiral arms spinning out from it. These arms glow with the light of the stars in them. The Andromeda galaxy has over 300 billion stars.

Island universes —
another name for galaxies?

In 1755, a German philosopher, Immanuel Kant, wondered about certain foggy patches astronomers could see through the telescope. Might these patches be very distant collections of stars? He thought they were, and he called them island universes . No one took his thoughts seriously. They thought the foggy patches were just clouds of dust and gas fairly close to us. It took astronomers nearly 200 years to work out the nature of those foggy patches (called nebulae), and then they found out that Kant had been right all those years ago. What he called island universes, we now call galaxies.

15

A large cluster of galaxies in the Coma Berenices constellation. This cluster contains about 1,000 galaxies.

NOAO

Clusters of Galaxies

A cluster like the Local Group isn't so unusual. Like stars, galaxies usually exist in clusters. Our Local Group is a rather small cluster of galaxies. Other, larger clusters exist, too. In the constellation Coma Berenices there is a cluster that contains nearly a thousand galaxies. This cluster is nearly 400 million light-years away. Nearer to us is a cluster of galaxies in the constellation Virgo that is made up of 2,500 galaxies. Recently, a still larger cluster containing over 25,000 galaxies was discovered. Our own Galaxy, huge though it is, is only one of many billions of galaxies in our Universe. We haven't counted them all. But we know they're out there!

The Virgo cluster of galaxies. This is an irregular cluster, which means that it is rather loose and not so tightly concentrated toward the cluster's centre.

NOAO

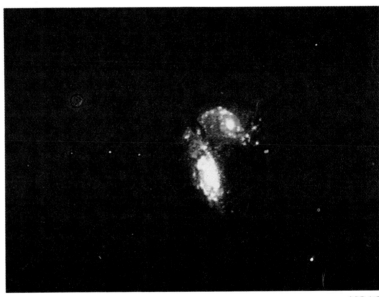

The 'Siamese Twins' galaxies. They are called this because of the apparent contact between the two galactic disks.

NOAO

Another case of the missing mass?

Clusters of galaxies hang together, held by gravity. However, the gravitational pull of the stars we can see in those clusters just isn't enough to keep them together and prevent them from drifting apart. The only way for astronomers to explain this is to suppose that there is more mass present that can't be seen because it is not produced by stars. The only trouble is that no one is sure what produces that mass. Very dim stars? Planets? Or mysterious objects like nothing we've ever seen? It may be yet another case of the 'missing mass' — but we don't know.

Two views of the Milky Way Galaxy: left, a face-on view, with the spirals whirling about the galactic centre; right, a view from the side, with the spirals forming a disk or plane intersected by the galactic centre. The centre bulges with older stars, while the spiral arms are the home of brightly-shining new stars.

Our Spiral Galaxy

Not all galaxies are shaped the same way. Many are elliptical, or oval-shaped. Others are spirals, with flat, round, swirly shapes.

Our Milky Way is a spiral galaxy. Its outer regions are made up of long, curved lines of stars, called spiral arms. These curve into the central part of the Galaxy. This is the part that contains about 90 percent of all the stars in the Galaxy. Astronomers trace the spiral arms by following the young giant blue stars they contain. The Orion arm contains our Sun. It is the third of the Milky Way's four arms. The Centaurus and Sagittarius arms are closer to the centre of the Galaxy, and the Perseus arm is farther from the centre. All the stars in these arms move around the centre of the Galaxy. The Sun circles the centre once every 230 million years.

Here is a closer look at the illustrations on the opposite page, with some of the major parts of our Galaxy marked. See how the Orion arm is formed. Many people call it a <u>spur</u> rather than an arm.

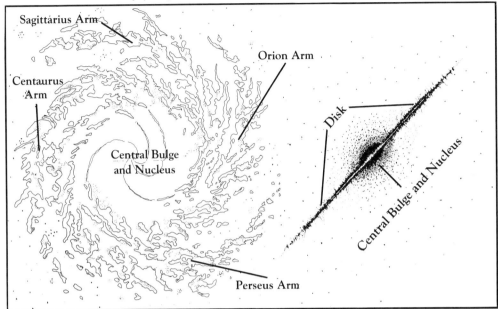

Sheri Gibbs

© Sally Bensusen 1987

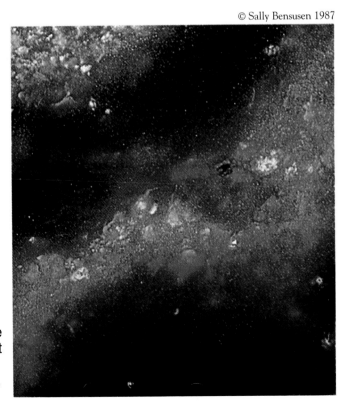

...use our Solar system is on the edge of the Orion arm, we get ...ctacular view of the next ...most arm, the Sagittarius arm, ...n here in this illustration.

Other Spirals

Our Milky Way is a wonderful sight. But of course we can only see it from the inside, so we don't get a good overall view of it. However, we can see other galaxies, and some of them have beautiful spiral shapes, especially if we happen to see them face-on. The so-called Whirlpool galaxy is perhaps the most beautiful of them, for it really looks like a whirlpool. The Andromeda galaxy is seen at a slant, but you can still make out the arms. Some, like the Sombrero galaxy, are seen edge-on, and you can usually make out a line of dust clouds along the edge of the rim. Almost every spiral galaxy is beautiful in its own way.

Anglo-Australian Telescope Board, David Malin

Above: a spiral galaxy visible from Earth's Southern Hemisphere.

Halton C. Arp

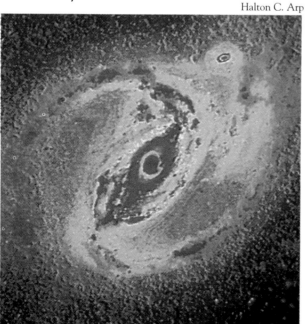

Left: a false-colour picture of a galaxy with a possible black hole at its centre.

Right: M81— not a catchy name, but this is one of the most visible and familiar of the spiral galaxies. The computer-enhanced colour shows the younger stars that make up the arms as blue and the older stars of the disk as orange.

Lower left: the Sombrero galaxy. What a beauty! Any guesses as to why they call it 'Sombrero'?

Lower right: dust clouds across the plane of a spiral galaxy blotting out the light of background stars. The next time you go barefoot in your garden or in a park, think about how every particle of dirt under your feet once floated in space in clouds like these.

Halton C. Arp

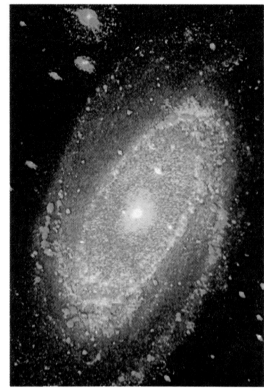

USNO

NOAO

21

Ellipticals

Many galaxies do not have spiral structures. They seem to be made up just of centres, without arms. These armless galaxies are called elliptical galaxies because their outlines are usually elliptical. Most of them are rather dim, and they are not as spectacular as the spiral galaxies. Some elliptical galaxies are giants, however. Large clusters of galaxies are often made up of ellipticals, and the largest contain up to a hundred times as many stars as our Galaxy does! Elliptical galaxies, like the centres of spiral galaxies, are made up mostly of old, not very large stars.

NOAO

Centaurus A is one of the brightest and largest of the known galaxies. Scientists think that giant explosions involving millions of stars are occurring at the galaxy's nucleus. The matter ejected by these explosions would show up as the dark band across the galactic disk.

© Garret Moore

A variety of galaxy types as seen from an imaginary planet. Can you pick out the ellipticals and spirals?

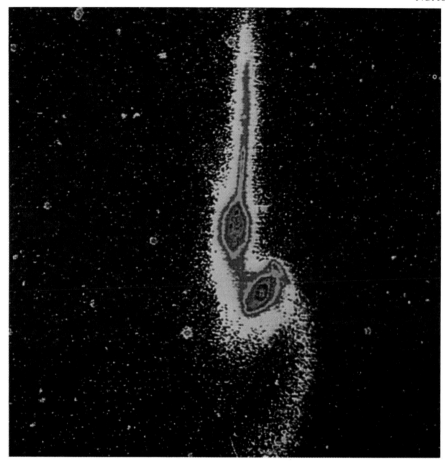

Some galaxies interact or even collide. This computer-enhanced picture shows streams of gas created by the interaction of two 'mouse' galaxies.

How many stars did you say?

The average galaxy may have about 100,000,000,000 (a hundred billion) stars. Some giant elliptical galaxies have a hundred times that number. On the other hand, there are many dwarf galaxies with only a tenth that number. Still — if a hundred billion stars per galaxy is the average and if there are about a hundred billion galaxies, as most astronomers think, the total number of stars in the Universe is about 10,000,000,000,000,000,000,000 stars!

Exploding Galaxies

There are small regions at the centre of galaxies that give off powerful radiations — light, radio waves, x-rays, and so on. Even our own Galaxy has a very active centre. Astronomers suspect there are black holes at the galactic centres. These black holes are small objects with a mass equal to that of millions of stars. They have strong gravity — so strong that they can swallow additional stars and even light itself. Some galaxies have centres so active that they seem to be exploding, sending out jets of matter from the centre and emitting vast radiation. Our own Galaxy, fortunately, is fairly well-behaved!

Halton C. Arp

M82, a galaxy that appears to be exploding. The main body of redder stars looks shattered, and gas seems to be shooting out of the centre in a bluish cone of radiation.

A swirling spiral galaxy, shown in the smaller picture at lower right. The large picture gives a close-up of the galaxy's active centre. Here we see a binary, or double, star system to stay clear of! A black hole is drawing glowing gases from its stellar companion — a star like our Sun — into the black hole's swirling disk. Though it is called a hole, this black hole is far from empty. In fact, it is incredibly dense, with the mass of millions of stars, and with gravity so strong that not even light can escape. The black hole's twin jets expel radiation and excess matter.

Very — make that <u>incredibly</u> — distant galaxies

Galaxies that are very far away — say, over a billion light-years away — are too dim to be seen. However, a few galaxies that are that far away have very active centres. Those centres blaze so brightly they can be seen even when the rest of the galaxy can't be. Astronomers detected certain dim stars they thought belonged to our own Galaxy. Imagine the surprise when analysis of their light showed they were the centres of <u>incredibly</u> distant galaxies that only looked like dim stars because of their great distance. Those far-off active centres are now called quasars — and they are among the most distant objects ever seen in the Universe!

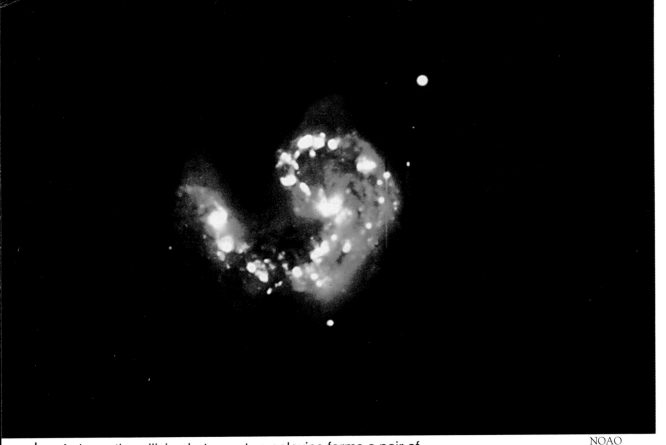

A dramatic collision between two galaxies forms a pair of 'antennae' or 'rattail' galaxies, 90 million light-years from Earth. This encounter began 500 million years ago.

Galaxies in Collision

It's a big Universe. Galaxies in clusters do move about, however, and some galaxies collide with each other as a result. In most cases, stars are so far apart that they pass each other harmlessly, but colliding galaxies <u>can</u> have an effect on one another. The breaking up of dust clouds, for example, can produce a lot of radiation. And if galaxies collide head on, they sometimes remain together. In fact, the giant galaxies in some clusters may be as large as they are because they have swallowed others. The giants are sometimes called 'cannibal galaxies' for that reason. We can't be sure, but after four billion years or so our Galaxy may collide with the Andromeda galaxy. Who knows what the effects would be? For now, at least, we're safe in our corner of the Milky Way!

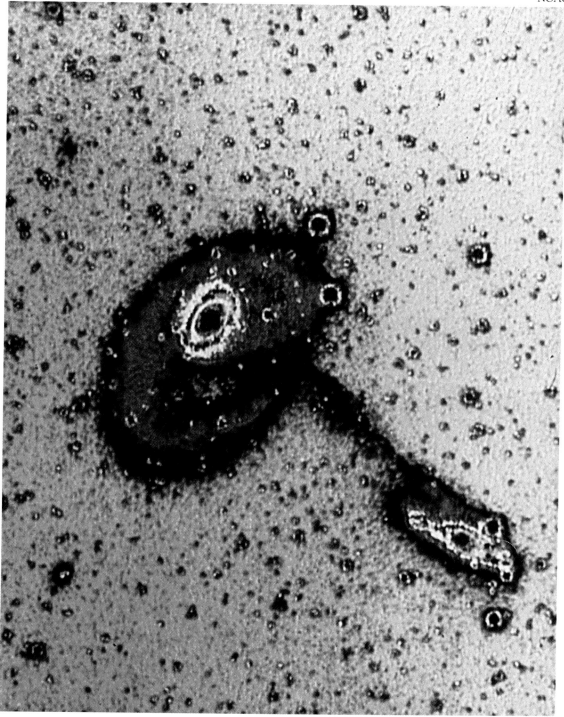

A computer-enhanced picture of two 'toadstool' galaxies in collision.
These galaxies are connected by a bridge of gas. The bridge is lit by
bright, young stars.

Fact File: Our Milky Way and Other Galaxies

Constellations: Mapping the Milky Way

When you look up into the Milky Way, you see stars — thousands of them at once. Some of these stars stand out from the rest, and for centuries people have thought that stars form patterns or even pictures in the sky. These are called constellations. Most constellations are named after objects, creatures, and gods from ancient mythology.

Because of Earth's tilt, not all constellations are visible from any one spot on Earth. Also, the stars that make up any one constellation are not actually close together. They just look that way from where we stand. So constellations don't really tell us much about the physical relationship of one star to another. But constellations do test our imaginations, and they help us find stars, planets, and other objects in the sky. They are our map of the Milky Way.

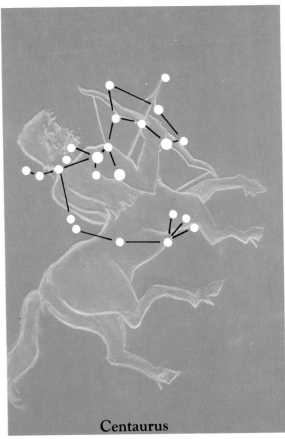

Centaurus

The spiral arms of our Galaxy have been Centaurus, the Centaur, half man and Perseus, the hero who killed Medusa,

The Galaxy of Numbers: How Big Is a Billion?

How big is a billion? You probably know what a **two**-storey school building looks like. If you have ever watched a game of cricket, you can imagine about **20 m** (**66** feet) — the distance between each wicket. And you can probably figure out about how long **two** hours seems — or **24** hours, which makes **one** day.

But what if we say that light travels nearly **9.6 trillion** km, or **six trillion** miles in a year? That's quite another story. And what if, to make matters worse, we say that the Milky Way Galaxy is **100,000** light-years across? This would mean that the Milky Way is **100,000** times **9.6 trillion** km across! Can you imagine numbers that high? Who could?

Reading about galaxies means reading about time and space — usually huge amounts of time and space. It therefore means seeing numbers so large that it may be impossible to understand exactly what they mean. How big <u>is</u> a billion? We'll probably never really know from anything we do in our day-to-day lives.

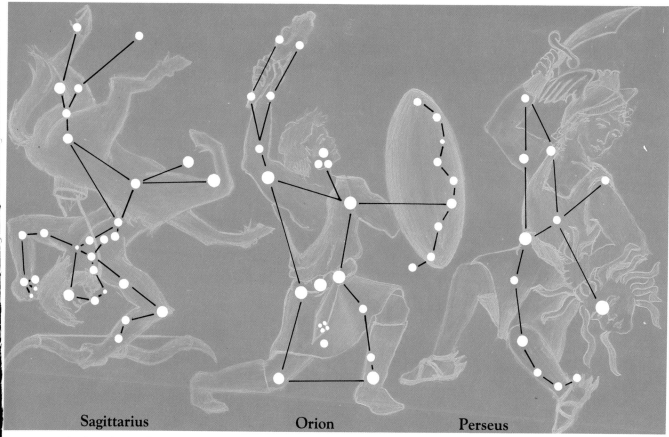

Sagittarius   **Orion**   **Perseus**

©Laurie Shock 1988

named after these four constellations. Left to right, they are half horse; Sagittarius, the Archer; Orion, the Hunter; and the terrible creature that turned people to stone.

But perhaps we can imagine, especially if we start with numbers that are small:

- The 16 sheets of paper that make up this book total about two millimetres, or 1/16 inch, in thickness.
- A **million** sheets of paper would be as high as a 32-storey building.
- A **billion** sheets of paper would be over 90 km (55 miles) tall — 10 times taller than Mt. Everest!
- A **trillion** sheets would tower more than 95,000 km (59,000 miles) above Earth — more than one-quarter of the way to the Moon!

- If we count 60 seconds to a minute, and 60 minutes to an hour, then 86,400 seconds makes up a 24-hour day.
- A **million** seconds is 12 days.
- A **billion** seconds is more than 31 years.
- A **trillion** seconds adds up to 300 centuries. That's 30,000 years. Thirty thousand years ago, many of our ancestors were still living in caves!

No matter how you look at it, one thing is certain: The numbers may be small to begin with, but jumping from a **million** to a **billion** to a **trillion** is no small matter. How quickly small numbers become astronomical!

29

More Books About the Milky Way and Other Galaxies

Here are more books that contain information about galaxies. If you are interested in them, check your library or bookshop.

Book of Stars and Planets. Maynard (Usborne)
Colour Library of Science: Astronomy. P L Brown (Orbis Pub.)
Exploring The Earth and The Cosmos. Asimov (Penguin)
How Was the Universe Born? Asimov (Gareth Stevens)
Outer Space. Asimov (Longman)
The Solar System. Ryan/Pesek (Penguin)
Space Facts and Lists. Reid (Usborne)
The Stars Above. Moore (Jarrold, Norwich)

Places to Visit

You can explore our Milky Way and other galaxies in the Universe without leaving Earth. Here are some museums and centres where you can find a variety of space exhibits.

The London Planetarium
London

The Science Museum
London

The Royal Observatory
Edinburgh, Scotland

Armagh Planetarium
Armagh, Northern Ireland

The Royal Greenwich Observatory
Herstmonceux Castle, Hailsham, Sussex

For More Information About the Milky Way and Other Galaxies

Here are some places you can write away to for more information about our Milky Way and other galaxies. Be sure to tell them exactly what you want to know about or see. Remember to include your full name, address, and age so they can write back to you.

For photography of galaxies:
Science Museum Library
Photo Orders Service
South Kensington
London SW7 2DE

For information about astronomy:
Junior Astronomical Society
36 Sandown Way
Greenham
Newbury RG14 7SD

For catalogues of slides, posters, and other astronomy materials:
Earth and Sky
21A West End
Hebden Bridge
West Yorkshire   HX7 8UQ

Spaceprints
17A High Street
Norton
Stockton-on-Tees, Cleveland

Glossary

accretion disk: a ring of interstellar matter surrounding a star or other object, such as a black hole, with an intense gravitational field.

astronomers: people who study the various bodies of the Universe.

billion: in the USA, and in this book, the number represented by 1 followed by nine zeroes — 1,000,000,000. In other countries, including Britain, this number is called 'a thousand million'. In these countries, one billion would then be represented by 1 followed by *12* zeroes — 1,000,000,000,000: a million million, which is called a 'trillion' in this book.

black hole: a massive (small but very heavy) object — usually a collapsed star — so tightly packed that not even light can escape the force of its gravity.

'cannibal' galaxies: giant galaxies that have collided with and 'swallowed' other galaxies.

constellation: a grouping of stars in the sky that seem to trace out a familiar figure or symbol. Constellations are named after that which they are thought to resemble.

double stars: stars which circle each other.

elliptical: oval-shaped.

galaxy: any of the billions of large groupings of stars, gas and dust that exist in the Universe. Our Galaxy is known as the Milky Way.

globular star clusters: ball-shaped groupings of closely-packed stars. These clusters are smaller than galaxies, although there may be hundreds of thousands of stars in each ball.

Kant, Immanuel: a German philosopher who was the first to think that patches seen through telescopes might be what he called 'island universes'. We now call them galaxies.

light-year: the distance which light travels in one year — nearly 9.6 trillion km, or six trillion miles.

Milky Way: the name of our Galaxy. From Earth's position in the Galaxy, the Milky Way looks like a river of stars in the night sky.

Orion arm: that part of our Galaxy where Earth is located.

spiral arms: long, curved lines of stars. Our Milky Way consists of such spiral arms.

Sun: our star and provider of the energy which makes life possible on Earth.

ten sextillion: in this book, 10,000,000,000,000,000,000,000; also 10 billion trillion. This is one estimate of the total number of stars in the Universe. (See **billion** above.)

Universe: everything that we know exists and believe may exist.

Index

The publishers wish to thank the following for permission to reproduce copyright material: front cover, pp. 6, 8,
10, © Julian Baum 1988; p. 22 (lower), © Garret Moore; pp. 4-5, © Frank Zullo 1985; p. 7 (right), © Frank Zullo
1986; pp. 7 (right), 20 (upper), Anglo-Australian Telescope Board, David Malin; pp. 9, 11 (upper), 12-13, 13, 14-
15, 16, 17 (both), 21 (lower left), 22 (upper), 23, 26, 27, National Optical Astronomy Observatories; p. 11 (lower),
© John Foster; pp. 18-19, © Lynette Cook 1987; p. 19 (upper), Sheri Gibbs; p. 19 (lower), © Sally Bensusen
1987; p. 25, © Sally Bensusen 1988; pp. 20 (lower), 21 (upper), 24, courtesy of Halton C. Arp; p. 21 (lower right),
Official US Naval Photograph, US Naval Observatory; pp. 28-29 (all), © Laurie Shock 1988.